SU

& EARTH

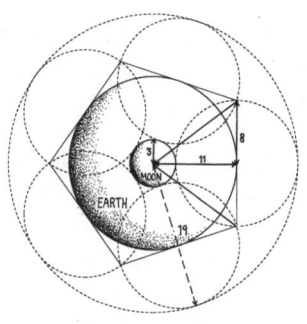

Originally published in Wales by Wooden Books Ltd. in 1999; first
published in the United States of America in 2001 by
Walker Publishing Company, Inc.

Published simultaneously in Canada by Fitzhenry and Whiteside,
Markham, Ontario L3R 4T8

Printed on recycled paper.

Library of Congress Cataloging-in-Publication Data

Heath, Robin.
Sun, moon, & earth / written and illustrated by Robin Heath.
p.   cm.
Originally published: Wales : Wooden Books, 1999.
ISBN 0-8027-1381-5 (alk. paper)
1. Archaeoastronomy.   I. Title: Sun, moon, and earth.   II. Title.
GN799.A8 H47   2001
520--dc21   2001026012

Printed in the United States of America

2 4 6 8 10 9 7 5 3 1

SUN, MOON, & EARTH

written and illustrated by
Robin Heath

Walker & Company
New York

For all the new children

Special thanks go to my brother, Richard, for being a vital part of the early research. Many thanks to John Martineau.

Note: Some descriptions in this book assume you are living in the northern hemisphere.

The title page shows the Venus of Laussal, circa 18,000 B.C. A clear message survives the aeons, confirming ancient human knowledge of the link between the Moon and the human reproductive cycle. Thirteen notches on a crescent horn link astronomy with human culture.

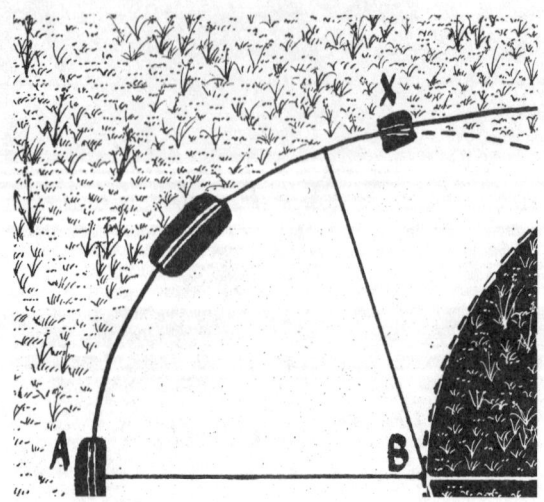

Above: Bar Brook, Derbyshire. A typical type-B flattened stone circle, 4,500 years old and revealing a subtle cosmology and metrology.

CONTENTS

Tarot card numbers eighteen and nineteen—the Moon and the Sun—framed by a playing card set: 4 suits of 13 cards, representing the 52 weeks in the year. Adding up the numerical values of each suit (the sum of 1 to 13) yields 91, the number of days per season. All four seasons then total 364 days, the joker completing the year at 365. An ancient aide memoire *for the calendar.*

INTRODUCTION

In the frenzied attempt to better understand and control the material universe, our present culture has strayed so far from simplicity and beauty that we are often startled or afraid when it reveals itself.

The modern system of ideas we call *science* has dispensed with the poetic and broadly fails to see the subtle connecting strands of meaning woven into the web of life. In addition, science is today shackled to commerce, also blind to such things, and thus we have "the blind leading the blind." As if that wasn't enough, today's high priests of science also inform us which interpretations of the cosmos are valid and which are not. Some questions are just not to be asked anymore, let alone answered.

This little book reveals a poetic cosmology that lies within the cycles of the Sun and Moon, as seen from Earth. It is found to be supremely rational, so simple and elegant that no priestly intermediary is needed to interpret, censor, or intervene.

All the mathematics given here may be verified by those of little faith with a simple calculator and the mind of an inquisitive teenager.

St. Dogmaels, 2001

SEARCHING FOR PATTERNS
finding order in the cosmos

The first systematic observations of the Sun and Moon are shrouded in the mists of prehistory. Scored bones from 40,000 B.C. (*below*) display lunar number cycles while the famed *Venus of Laussal* (*title page*) links the Moon with the number thirteen.

Repeated cycles such as full moons, eclipses, and planetary conjunctions revealed a cosmology to ancient astronomers that was both numerical *and* geometrical, and which imbued creation with order and meaning—"God is a geometer." The delphic adage "as above, so below" suggests that cosmic patterns are reflected in earthly life, becoming a source of revelatory information.

The Great Pyramid (2480 B.C.) epitomizes this approach. Built to the points of the compass, with passageways aligned to stars, its base and height fit the "squared circle" of Earth and Moon.

This archaic approach to cosmology is today discarded as worthless, and has been replaced by modern astronomy. Yet most people know almost nothing about the Sun, Moon, and Earth system, despite our total dependence on its rhythms. This book will gently put that right, and reinvoke something of the spirit of the old sciences.

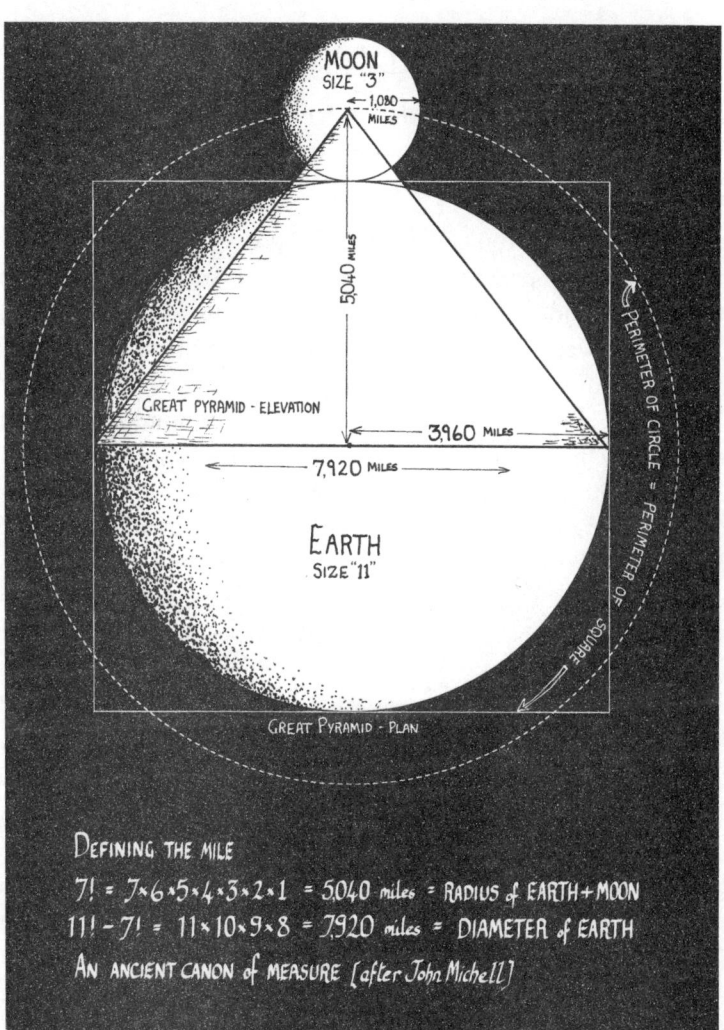

MOON
SIZE "3"

← 1,080 →
MILES

5,040 MILES

GREAT PYRAMID - ELEVATION

3,960 MILES

7,920 MILES

EARTH
SIZE "11"

PERIMETER OF CIRCLE = PERIMETER OF

PERIMETER OF SQUARE

GREAT PYRAMID - PLAN

DEFINING THE MILE

7! = 7×6×5×4×3×2×1 = 5,040 miles = RADIUS of EARTH + MOON

11! ÷ 7! = 11×10×9×8 = 7,920 miles = DIAMETER of EARTH

AN ANCIENT CANON of MEASURE [after John Michell]

SOME EARLY SOLUTIONS
from megaliths to the Maya

Skywatching is an ancient art. Stone circles date from 3000 B.C., aligned megaliths even earlier. The Egyptians were using accurate surveying and a precise metrology for both sky and Earth. The Great Pyramid enshrined its date of construction through astronomical alignments to fixed stars. The Sumerians recorded astral cycles from 2200 B.C. and later defined the 24-hour day and 360-degree circle. Chaldean and Chinese astronomers knew of the Saros eclipse cycle (*page 28*). Various calendars were in use.

From 600 B.C., the Greeks inherited this ancient wisdom. Eratosthenes measured the size of the Earth and Eudoxus devised a solution for the complex motion of the Moon. In the fourth century B.C., the nineteen-year cycle of Sun and Moon was described by Meton. The Romans gave us our modern calendar in 45 B.C.

When the Empire collapsed around A.D. 500 the Arab world kept the torch of learning burning as Europe sank into the Dark Ages. Following the Crusades this material returned, seeding the Renaissance in Europe. Copernicus showed that the Earth orbited the Sun, while Galileo's telescope revealed moons orbiting other planets. Kepler published the three laws of planetary motion in the early seventeenth century, when Newton used data about the Moon to quantify his universal laws of motion and gravity in 1687, thereby spawning our modern world. In the next century, Harrison's chronometer greatly improved timekeeping and navigation.

El Castillo, Chichén Itzá

4 flights of 91 steps totaling 364
plus the high altar—365.

EGYPTIAN BAY

MERKHET

~ Early Surveying Instruments from Egypt ~

These instruments belonged to an "Hour Priest"
of the twenty-sixth Dynasty, circa 1000 B.C.

THE SUN
the day and the year

Each day, the Sun appears to rise from an easterly direction, traces a clockwise arc across the heavens, and then sets toward the west, disappearing for the dark time we call night. This cycle repeats perpetually; it is the diurnal rhythm, called, more simply, a day.

Today we are taught that what we see is caused by the daily rotation of a spherical Earth orbiting the Sun. Thereafter, like The Fool on the Hill, we "see the Sun going down, while the eyes in our head see the Earth spinning round." Each day, the Sun appears to move about a degree counterclockwise (eastward) with respect to the fixed stars. Thus the solar day, to which we set our clocks, exceeds the *sidereal* (star) day by 3 minutes and 56 seconds.

The axial tilt of the Earth (*page 9*) causes the Sun to rise and set each day at different positions on the horizon. Only at the summer and winter *solstices* (see page 8) is this daily change in the Sun's rise and set positions reduced to zero, at their extreme *standstill* positions. Subsequent sunrises and sunsets gradually reverse back along the horizon, the span being dependent on the latitude of the observer (*opposite, bottom*). This is the rhythm of a year.

The Earth's solar orbital period is 365.242199 days. Our "solar" calendar of 365 days keeps pace by adding regular leap-year days, one every four years (except once every four hundred years), and the odd second or two.

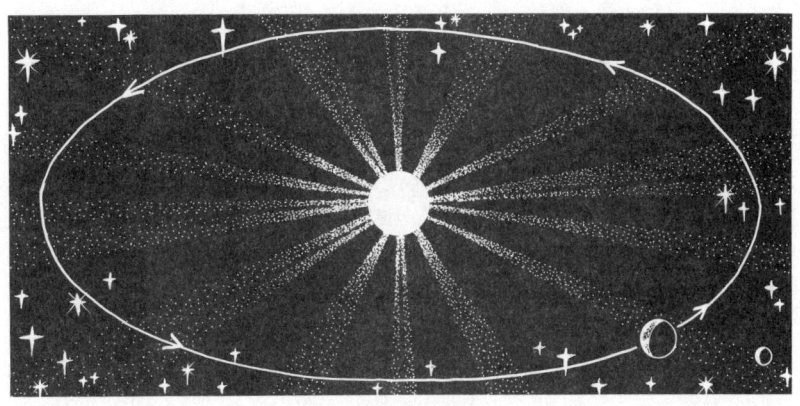

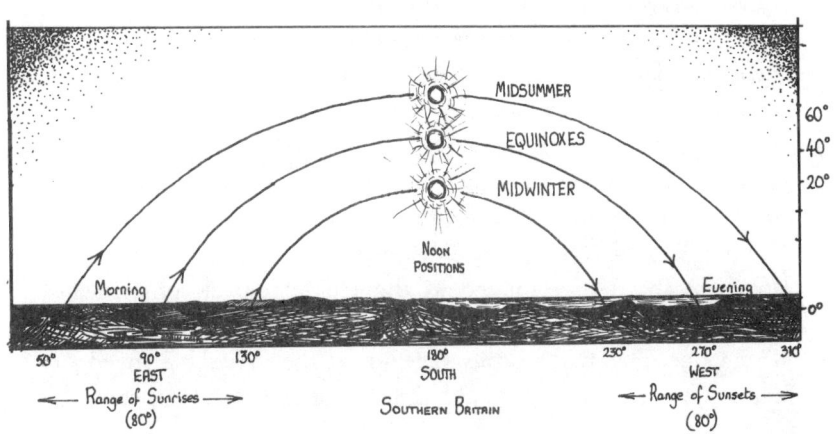

MIDSUMMER

EQUINOXES

MIDWINTER

NOON
POSITIONS

Morning

Evening

60°

40°

20°

0°

50°   90°   130°   180°   230°   270°   310°
EAST   SOUTH   WEST
← Range of Sunrises →   ← Range of Sunsets →
(80°)   SOUTHERN BRITAIN   (80°)

SOLSTICES AND EQUINOXES
the four natural divisions of the year

In between the two solstices, the longest and shortest days in the year (normally June 21 and December 22), lie the two equinoctial periods in the spring and the autumn. The *equinoxes* (March 21 and September 23) deliver equal lengths of day and night everywhere on the planet, with the Sun rising exactly due east and setting exactly due west, on a level horizon.

These equinoctial dates are accompanied by the maximum rate of change in the length of the day. In temperate latitudes, this creates the impression that the year is divided into two distinct halves, a light, warm summer half and a dark, cold winter half. During the summer half the Sun rises and sets north of an east-west line; in the winter half always south of it.

The solstices and equinoxes naturally divide the year into four quarters, defining the four seasons. Each season is 91 days in length (*see page 5 and opposite page 1*), caused by the tilt of the Earth on its own axis (currently $23\frac{1}{2}°$ with respect to its orbital plane). This angle may be constructed using a right triangle, base 13 and height 30, or more approximately 3 and 7.

The "cross-quarter" days, halfway between equinoxes and solstices, are still celebrated as the Celtic festivals of *Samhain* (November), *Imbolc* (February), *Beltane* (May Day), and *Lughnasadh* (August). The Earth orbits the Sun at the incredible speed of 66,666 miles per hour and at a distance of 108 solar diameters.

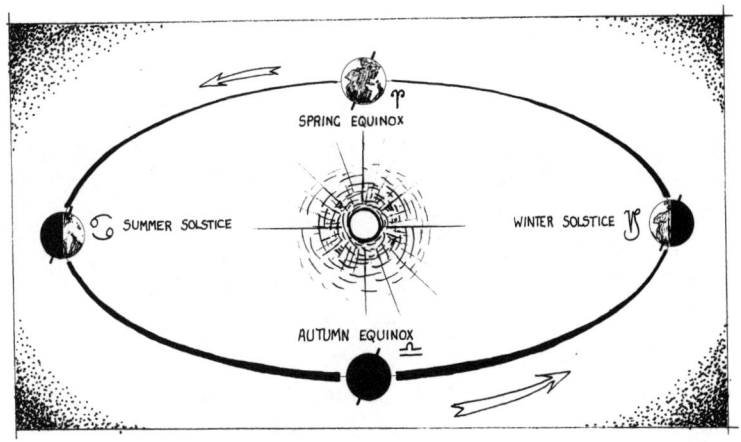

SPRING EQUINOX

SUMMER SOLSTICE

WINTER SOLSTICE

AUTUMN EQUINOX

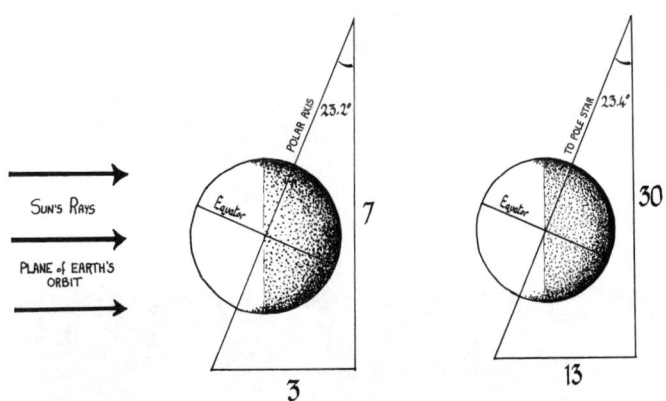

SUN'S RAYS

PLANE of EARTH'S ORBIT

POLAR AXIS 23.2°

Equator

TO POLE STAR 23.4°

Equator

9

THE MOON
the goddess of the night

Although apparently lifeless, the Moon greatly affects life on Earth. The fluctuating monthly rhythm of reflected light, the twice daily ebb and flow of the tides, and many natural cycles are all essentially locked into the lunar phases as, uniquely, is the reproductive cycle of humankind. The Moon is associated with women and the number 13, perhaps because the Moon moves 13 degrees a day and orbits the Earth 13 times in one year. People see a man in the moon or sometimes a hare, owl, swan, or lady.

At an average distance of 240,000 miles, the Moon is our nearest neighbor. Its radius is 1,080 miles compared with that of the Earth at 3,960 miles, a ratio of 3:11. However, the Moon is not spherical, and the Earth's gravity always pulls the larger hemisphere toward us. The Moon thus has its "dark side," which we never see, but which paradoxically becomes fully lit each new moon.

The Moon's orbital plane is tilted to that of the Earth (*below*). Periodically, this enables eclipses to occur and, at higher latitudes, every 18.618 years, causes wild monthly fluctuations in the altitude of the Moon, and a maximum angular range of rising and settings.

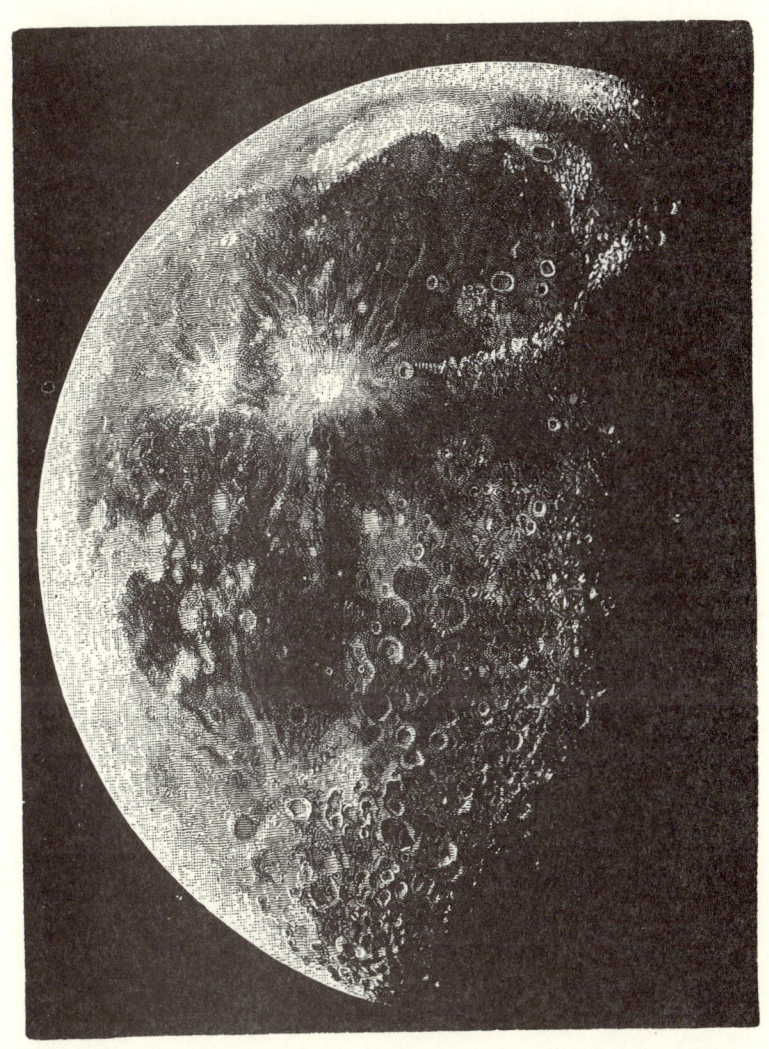

THE MOON'S TWO RHYTHMS
sidereal and synodic months

Observe the Moon for a short while and you will discover one of its rhythms — it moves past the fixed stars quite rapidly, taking about an hour to cover the distance of its own diameter. In one day, it covers 13 degrees, thereby taking slightly less than 28 days to return to the same stars. This is the *sidereal* month of 27.322 days, or 27 days, 7 hours, 43 minutes, and 11.51 seconds (approximately $27\frac{18}{56}$ days).

Three principle rotations are now defined: the *day* (Sun-Earth), the *sidereal month* (Moon-Earth-stars) and the *year* (Sun-Earth-stars). All three are shown opposite. But there is a fourth rotation, the lunar phases or *lunation cycle*, the time between full moons, which is truly Sun, Moon, and Earth. Because the lunar phases are visible all over the Earth, the lunation cycle is the prime lunar rhythm. It is also called the *synodic month* or, more simply, the *lunar month*. It takes 29.53059 days to complete, or 29 days, 12 hours, 44 minutes, and 2.37 seconds (approximately $29\frac{43}{81}$ days).

The Earth completes about a thirteenth of its annual journey around the Sun during one sidereal month, thus the lunar phases have to "catch up," as shown in the illustration (*opposite, bottom*), this taking an extra 2.21 days over and above the sidereal month.

There are 13.368 sidereal months and 12.368 synodic months in the year. The fractional part is very close to seven-nineteenths.

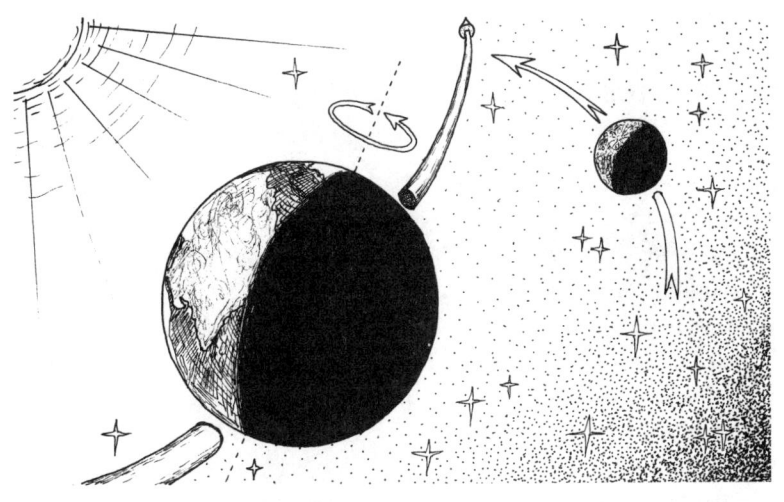

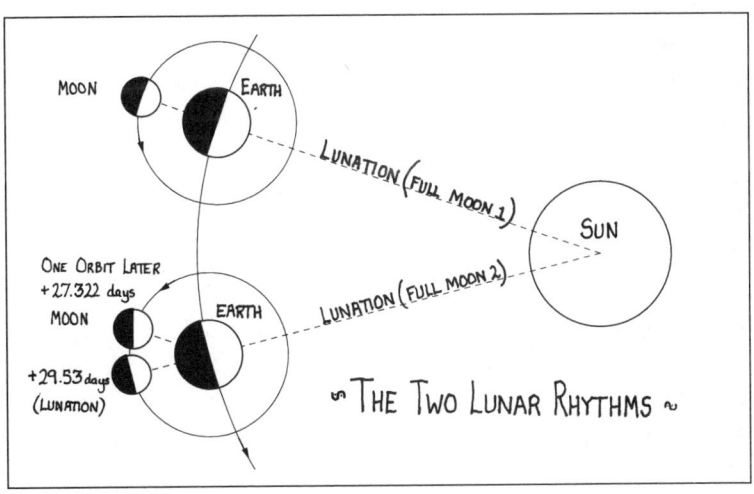

MOON   EARTH

LUNATION (FULL MOON 1)

SUN

ONE ORBIT LATER
+ 27.322 days
MOON

EARTH

LUNATION (FULL MOON 2)

+ 29.53 days
(LUNATION)

~ THE TWO LUNAR RHYTHMS ~

THE TICK-TOCK OF THE MOON
the lunation cycle or lunar month

The Moon begins her monthly phases acting as consort to the Sun, appearing as a sliver of silver to his left. This is the *new moon*, shaped like a reversed *C*. Each successive day finds the Moon belonging more and more to the night sky as the *waxing* phases increase the crescent to a *quarter*, *gibbous*, and then *full* moon, taking about thirteen days to complete.

Only when full does the Moon "escape" the Sun, becoming entirely nocturnal and reflecting the maximum silvery light down onto the night landscape. The *waning* cycle then progressively delivers the Moon back into the daytime skies as it leads the Sun, setting later and later in the day until, again after about thirteen days, it only becomes separate from the Sun just before dawn, glimpsed as a tiny *C*-shaped crescent to the Sun's right.

The Moon then disappears for about three days, lost in the light of the Sun at the new moon. This whole cycle is called the *lunation cycle*, lunar, or *synodic,* month; it is the time between full (or new) moons, and takes an average 29.53059 days to complete.

An inscribed 5,000-year-old curbstone at Knowth, in the Boyne valley, Ireland, displays what appears to be a representation of the lunation cycle. The impressive spiral correctly covers the three days of new moon, and 15 days later the full moon is marked ")(" within a 29-based motif. The "serpent" enclosed by this lunation motif has 30 turns, $29\frac{1}{2}$ being the average between the two.

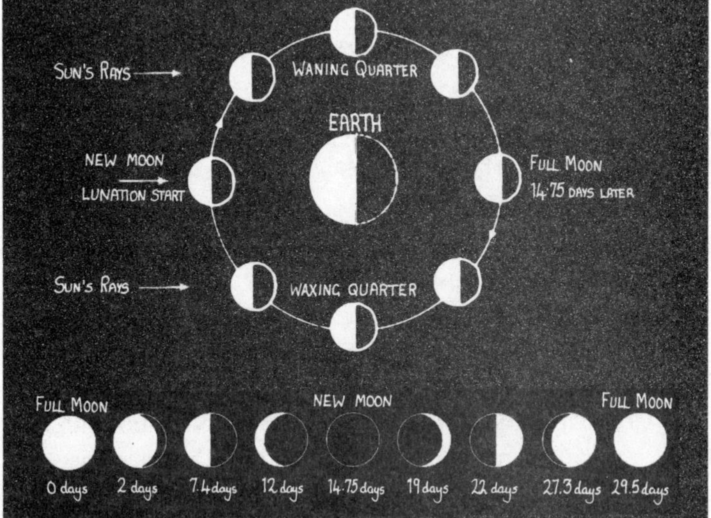

SUN'S RAYS → | WANING QUARTER

EARTH

NEW MOON
LUNATION START → | FULL MOON
14.75 DAYS LATER

SUN'S RAYS → | WAXING QUARTER

FULL MOON | NEW MOON | FULL MOON

0 days | 2 days | 7.4 days | 12 days | 14.75 days | 19 days | 22 days | 27.3 days | 29.5 days

~THE LUNATION CYCLE ~

KNOWTH K53

15

THE LUNAR DAY
tides of lunar time

Anyone who lives by the sea witnesses the inexorable silent power of the Moon, whose invisible claws draw the tides up and down the beach twice a day. Tides are not just limited to the oceans—the Earth's atmosphere above our heads and even its crust beneath our feet rise and fall to this lunar rhythm. The highest *spring* tides occur two or three days following a new or full moon. The low-range *neap* tides occur two days after a waxing or waning quarter moon.

The *lunar day* is the time between consecutive moonrises, the Moon rising an average of 52 minutes later each day. There are exactly two tides each lunar day, each one retarded by an average of 26 minutes every 12 hours. Tides are synchronized to the lunar day and therefore to the Moon's position in the sky. High tides will always occur at the same two positions of the Moon in the sky at any given location, these being opposite each other (one position is always beneath the horizon). A practical tidal indicator is shown on page 47.

There are $28\frac{1}{2}$ lunar days and therefore 57 tides (3 x 19) in each lunation cycle. In isolation tanks, human bodily rhythms eventually transfer from the solar to the lunar day.

The Earth and Moon form a huge dumbbell in space, with their center of rotation located about 1,000 miles beneath the Earth's surface (*shown as a small checkered circle opposite*).

The Moon's gravitational pull (M) lifts the oceans on the side of the Earth facing the Moon. On the opposite side, centrifugal force causes a similar effect (C), because the center of mass (and so of rotation) of the Earth-Moon system does not lie at the center of the Earth. The Sun also pulls at the oceans (S), and according to the phase of the Moon adds or subtracts to and from the height of the tide. "Spring" tides occur near full and new moons (upper), the lesser "neap" tides at the quarter-moons (lower). The monthly ratio between the heights of spring and neap tides is 8:3.

SUN, MOON, AND LANDSCAPE
the moon as mirror of the sun

Each month the Moon more or less copies the entire annual range of rises and falls undertaken by the Sun in a year.

The Moon rises highest in the sky each month when it is found near Betelgeuse, in Orion. Its most northerly risings and settings occur then. At extreme latitudes (above 60°) a midwinter full moon may become *circumpolar* and not set for a few days. The most southerly risings and settings occur when the Moon is found near Antares, in Scorpio. At extreme latitudes, for example in Finland or northern Canada, the full moon may not be visible during midsummer, especially above the arctic circle, where the Sun becomes circumpolar.

The full moon is brightest and highest at midwinter, copying the motion of the midsummer Sun. The midsummer full moon correspondingly behaves like the midwinter Sun, remaining low in the dusky sky. Thus the full moon mirrors the Sun at the opposite point in the calendar, and like a true mirror it fully reflects the Sun's light.

This reciprocation mysteriously extends into the numbers, for $1 \div Sun = Moon$, and $1 \div Moon = Sun$! $1 \div 365.242 = 0.0027379$, which in days is 3 minutes and 56 seconds, the difference between sidereal and solar days, while $1 \div 27.322 = 0.0366$, which in days is 52 minutes, the difference between lunar and solar days. It is fun to ask an astronomer why.

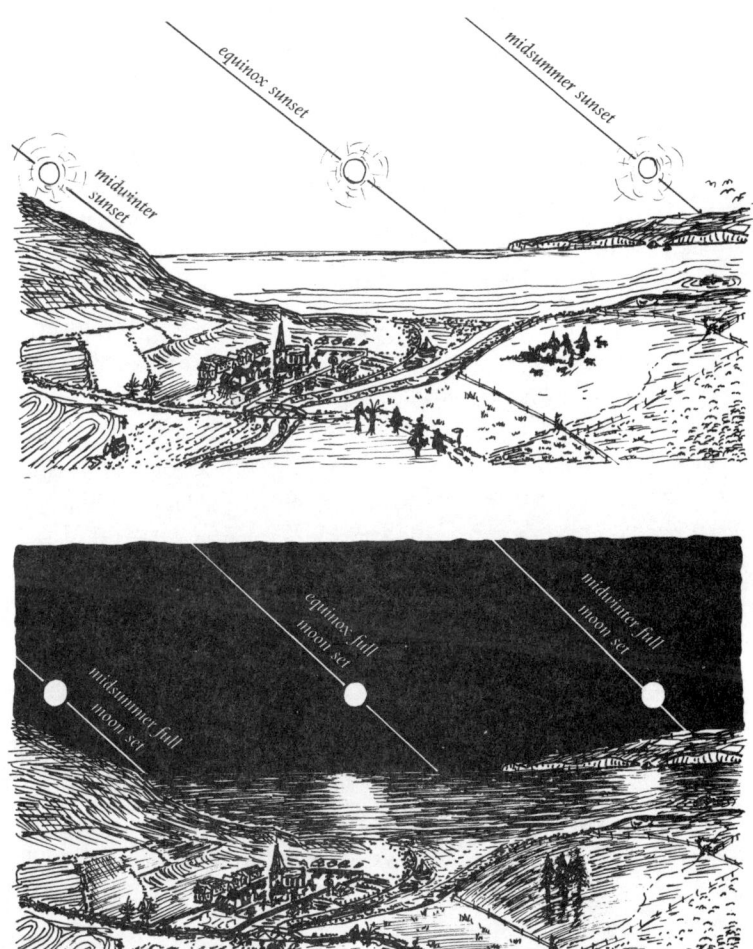

THE MOON'S NODES

the path of the moon crosses that of the sun

The orbit of the Moon is tilted with respect to that of the Earth by an angle of 5.14° (*see page 10*). The effect is that the Moon travels above the ecliptic (the apparent path of the Sun around the zodiac) for about half the sidereal month, and travels beneath it for the other half.

The two places where the Moon crosses the ecliptic each month are called the *lunar nodes,* and they always lie opposite each other. The two smaller illustrations opposite show these crossing points as observed from the Earth—but in truth, they are invisible! Eclipses only happen when a full or new moon occurs within $12\frac{1}{2}°$ or $18\frac{1}{2}°$, respectively, of the nodes; total eclipses when the alignment is almost exact. These are the *eclipse limits* for lunar and solar eclipses.

The axis of the nodes moves backward around the calendar, taking 18.618 years (6,800 days) to complete a circuit. It moves 19.618 days per year. To the ancients the nodes were thought of as the head and tail of a huge celestial dragon that swallowed the Moon or Sun during an eclipse. The nodal period is still known as the *Draconic year*.

The Sun meets a node every 173.3 days (an *eclipse season*); it meets a particular node after two of these periods have elapsed, this both defining and completing the *eclipse year* of 346.62 days.

Is it not the strangest thing that 346.62 = 18.618 x 18.618?

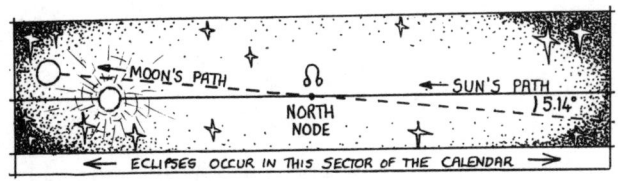

MOON'S PATH — SUN'S PATH →

♌ NORTH NODE

∠ 5.14°

← ECLIPSES OCCUR IN THIS SECTOR OF THE CALENDAR →

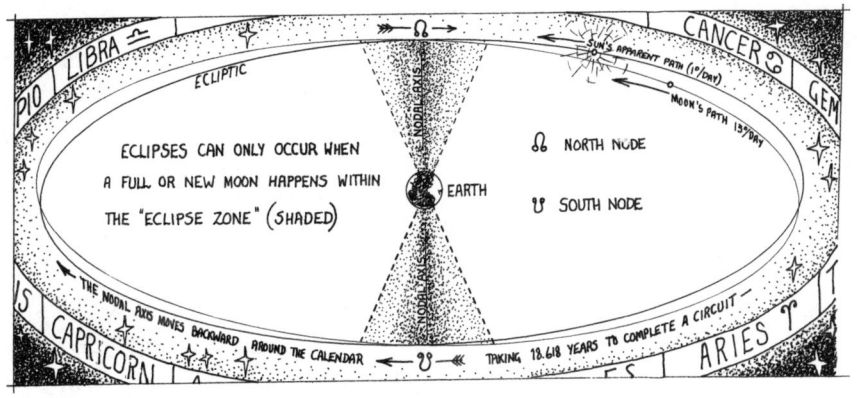

LIBRA ♎ CANCER ♋ GEM

SCORPIO ♏ ECLIPTIC

SUN'S APPARENT PATH (1°/DAY)

♌

MOON'S PATH 13°/DAY

ECLIPSES CAN ONLY OCCUR WHEN
A FULL OR NEW MOON HAPPENS WITHIN
THE "ECLIPSE ZONE" (SHADED)

EARTH

♌ NORTH NODE

☋ SOUTH NODE

CAPRICORN

THE NODAL AXIS MOVES BACKWARD AROUND THE CALENDAR ← ☋ ← TAKING 18.618 YEARS TO COMPLETE A CIRCUIT —

ARIES ♈

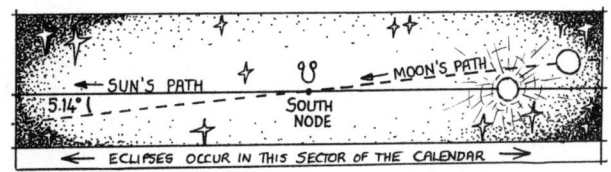

← SUN'S PATH ☋ MOON'S PATH ←

5.14° ∠

SOUTH NODE

← ECLIPSES OCCUR IN THIS SECTOR OF THE CALENDAR →

THE "BREATH" OF THE MOON
major and minor standstills

The monthly extreme northerly and southerly risings and settings of the Moon gently "breathe" in and out either side of the Sun's extreme solsticial positions, taking one nodal period to complete the "breath." This greatly alters the possible maximum rising and setting positions of the Moon each month with respect to the solsticial positions of the Sun. There are thus eight limiting "lunstice" positions, four for risings and four for settings (*opposite, top*).

The distance of these extreme positions of the lunstice from the solstice position is dependent on the latitude of a location. In southern Britain, they occur more than eight degrees of either side of the solstice positions (*opposite, bottom*). These extreme stations of the Sun and the Moon drew the attention of neolithic astronomers who made alignments of stones in their honor.

At the *major standstill*, the Moon describes her wildest monthly swings of rising and setting, gyrating to her highest- and lowest-ever paths across the sky, all within one sidereal month. At the *minor standstill*, 9.3 years later, the Moon calms down and the range always lies inside the solsticial positions.

In contrast to the Sun, Moon, and planets, the stars rise and set at exactly the same place along the horizon for hundreds of years irrespective of the season and time of day or night.

North
S
W
E
MIDSUMMER SUNSET
MIDSUMMER SUNRISE
MIDWINTER SUNSET
MIDWINTER SUNRISE
-RANGE of MOONSETS AT THE MAJOR STANDSTILL-
-MOONSET RANGE (MINOR STANDSTILL)-
-MOONRISE RANGE (MINOR STANDSTILL)-
-RANGE of MOONRISES AT THE MAJOR STANDSTILL-

LOCATION:
52°N 4°W40'
MIDWINTER FULL MOONRISE (MAJOR)
MIDSUMMER SUNRISE
MIDWINTER FULL MOONRISE (MINOR)
MIDSUMMER FULL MOONRISE (MINOR)
MIDWINTER SUNRISE
MIDSUMMER FULL MOONRISE (MAJOR)
Az48°
Az132°
NORTH
10°
10°
EAST
90°
10°
10°
SOUTH
MINOR STANDSTILL
MAJOR STANDSTILL

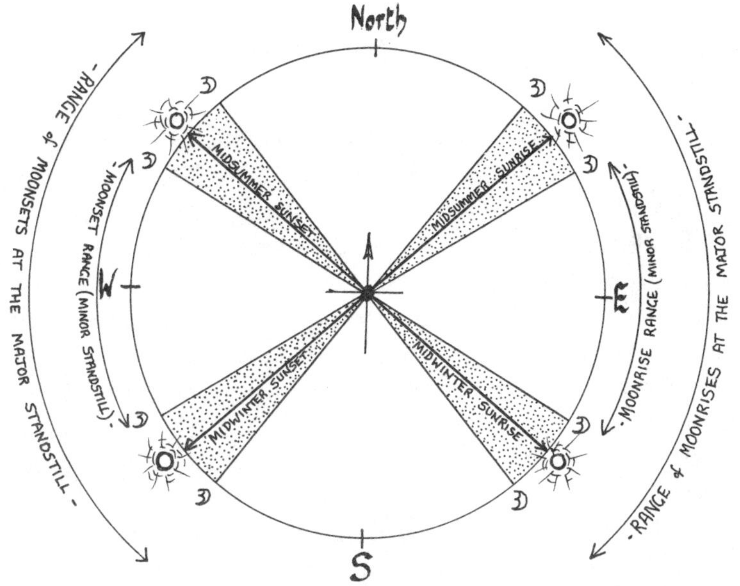

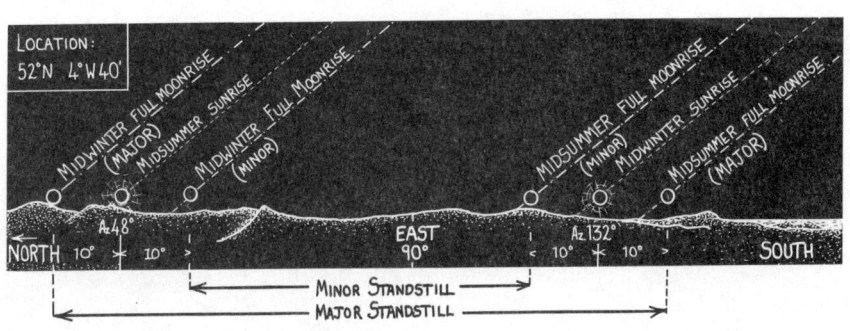

ECLIPSES
cosmic attention seeking

By what almighty coincidence do the disks of Sun and Moon appear the same size to us earthlings? The Sun is four hundred times larger than the Moon, yet four hundred times farther away. The distance of the Moon from the Earth is just over thirty Earth diameters. Total solar eclipses could never occur if the Moon's orbital distance was changed by just one Earth diameter.

Total solar eclipses instill an elemental awe in us, with a sudden brief reversal from light into darkness, after which "dawn" returns *from the west*, at over 2,000 miles per hour! Lunar eclipses are gentler and longer and simulate a whole lunation cycle in just a few hours.

Solar eclipses occur when the new moon passes directly between Sun and Earth. They can only be seen during the daytime, the area of totality tracing a narrow smudge of blackness across the Earth. Totality never lasts longer than seven minutes at any one location (*opposite, top*). During lunar eclipses the full moon passes from right to left through the Earth's shadow, its reflected light extinguished for several hours (*opposite, bottom*). Lunar eclipses are visible to all on the night side of the Earth.

As the angle between Sun and node increases, total eclipses decrease and become partial (*see page 30*). Beyond 18.5° no eclipse can occur. There can be up to seven eclipses in any one year, and solar eclipses are the more common, by the ratio $\sqrt{2}:1$.

LUNAR ECLIPSES
studies in light and darkness

A lunar eclipse is a striking phenomenon. The reflected light of the full moon greatly diminishes the light from the stars, and during the eclipse a curious and beautiful effect unfolds. As the full moon enters the Earth's shadow cone (*page 25, bottom*), the Moon's face darkens and the night sky radically alters its appearance, becoming brilliantly peppered with many more stars than were previously visible. This effect is also shown opposite, where a satisfying diagonal symmetry may also be seen.

During the period of totality, the Moon often takes on a remarkably beautiful coppery color within the starry firmament. Also stirring is the curve of the Earth's shadow as it draws across the lunar orb. It confirms that our planet is about three times larger than the Moon and spherical in shape.

Before 2500 B.C., Megalithic astronomers in northwestern Europe appear to have observed a tiny variation in the 5.14° tilt of the Moon's orbit ($\frac{1}{6}$° with period 173.3 days) in order to predict eclipses. Their observatories still exist, mainly in Scotland. What were our ancestors up to?

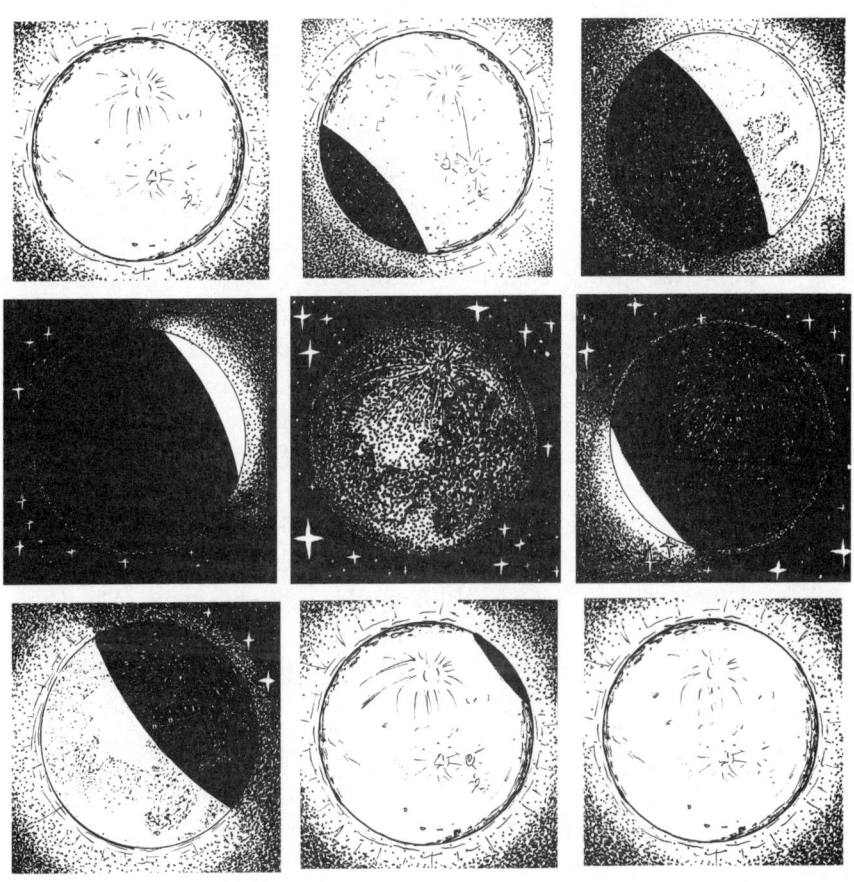

27

SOLAR ECLIPSES AND THE SAROS
the 18-year cycle of eclipses

There are three types of solar eclipse: *partial, annular,* and *total* (*all shown opposite*). These are produced by variations in orbital distances and nodal offset at *syzygy*, the barely pronounceable term for a full or new moon (Sun, Moon, and Earth in line).

Any particular eclipse is a member of a family, consecutive individuals of which display similar characteristics. A famous family is the *Saros cycle*, of 18 years and 11 days, 223 lunations or 19 eclipse years. A Saros cycle evolves and decays over about 1,300 years (solar eclipses) and 800 years (lunar eclipses). At any given time an average of 42 Saros families of solar eclipses, and 27 of lunar eclipses, are evolving, each delivering about 70 and 45 individuals respectively over its lifetime. The Saros was used by the ancient Chaldean astronomers to accurately predict eclipses.

The eclipse year (346.62 days) equals 11.738 lunations. Divide 19 eclipse years (the Saros) by 11.738 and you get 1.6186, almost *phi*, the *Divine Proportion*. Because the eclipse year is 18.618 x 18.618 days, the Saros may be written as 19 x 18.618 x 18.618 days (to 99.99%). Mysteriously, 19 lunations is *phi* eclipse years!

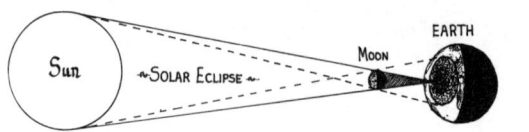

PARTIAL ECLIPSE

TOTAL ECLIPSE

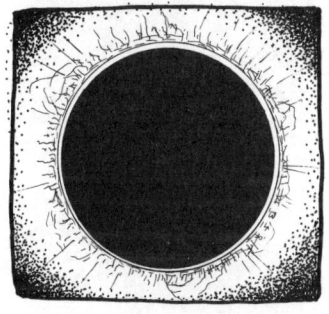

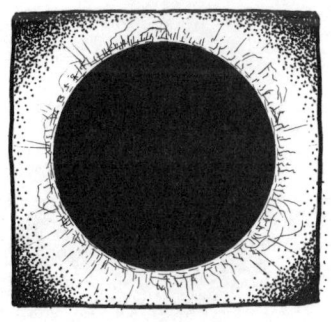

SAROS PATTERNS
a trilateral evolution

All Saros eclipse families first make an appearance at one of the poles, gradually evolving into equatorial regions before eventually dying out at the other pole (*below, bottom*). The time difference between the Saros and 19 eclipse years (0.46 days) causes each new member to be displaced about half a degree farther west with respect to the nodes. Thus the family takes about 36 Saros cycles (650 years) to reach the node (B) and thereafter departs from it in the same time, slowly dissipating (*below, top—every 7th Saros shown*).

The patterns made by these metamorphosing families of eclipses (solar or lunar) form threefold motifs on the Earth, due to the fact that each consecutive Saros period (223 lunations) is 6,585.321 days in length, the fractional component being about one-third of a day (or Earth rotation) out of alignment. The midpoints of the paths of totality for every third member may be joined up to reveal this threefold pattern. The result is curves called *exelegismos* (*opposite, dotted for solar and solid for lunar eclipses*).

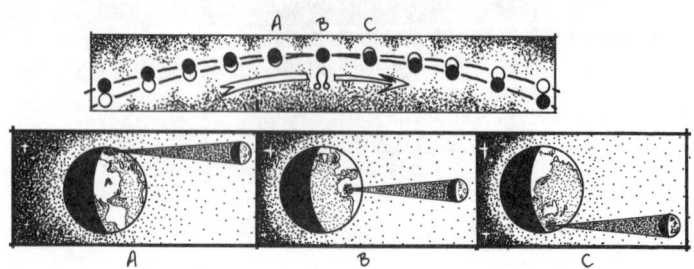

THE DANCE OF THE MOON
Ariadne's eight-fold web in the sky

The orbital distances of the Earth and Moon undergo periodic changes. This affects the duration and type of eclipses (*see pages 24–31*). When the Earth is nearest the Sun, strangely, in chilly January, it is said to be at *perihelion*; when farthest from the Sun it is said to be at *aphelion*.

Similarly, *perigee* occurs when the Moon is nearest the Earth, while *apogee* finds the Moon farthest from us. The line connecting these two points in the Moon's slightly elliptical orbit is called the line of *apsides*. This line or axis, the coming and going of the Moon, itself rotates, completing a cycle every 8.85 years, dividing the zodiac into eight (*opposite*) sections. A full moon at perigee appears 30 percent larger than at apogee.

The line of apsides moves counterclockwise around the zodiac by 40° 40' per year, while the nodal axis moves clockwise by 19° 20' per year (*see page 20*). The combined motion is thus 60° per year, a remarkable coincidence causing the nodes and apsides to rendezvous once more after 6 years (360°). Three of these meetings take 18 (6+6+6) years, coinciding almost exactly with the Saros cycle of 18.03 years. This is why eclipses that appear within consecutive Saros cycles are of the same type and duration.

In nautical almanacs and ephemerides, the position of the Moon is today predicted years in advance using a formula that contains over 1,500 separate factors.

A = Apogee   P = Perigee

33

THE 19-YEAR METONIC CYCLE
the marriage of Sun and Moon

After 19 years, and within two hours of exactitude, the Sun and Moon return to the same places in the sky on the same date. This important repeat cycle is named after Meton, a fourth-century B.C. Greek astronomer. It is an astonishingly accurate repeat cycle among several other contenders (*see table opposite; the outer stone circle at Avebury once comprised 99 megaliths*).

The first-century B.C. historian Diodorus suggested that the Celts knew of the 19-year cycle. We need not be surprised at his account, for the Celts inherited the culture of the stone-circle builders, and many fine circles, particularly in southwestern Britain, such as *Boscawen-un* in Cornwall (*shown opposite*), contain 19 stones. The bluestone horseshoe at Stonehenge comprised 19 slender dressed megaliths, brought from the Preseli Mountains of west Wales, 135 miles as the crow flies. Some weighed over 4 tons!

There are 12.368 lunations (full moons) in one solar year. The lunar year (12 lunations) falls short of the solar year by just under 11 days, which after 19 years accrues to 7 lunations, totaling 12 x 19 plus 7, or 235 lunations. From these numbers, the annual number of lunations may be found: $235 \div 19$ equals $12\frac{7}{19}$, this fraction revealing the underpinning astronomy.

Nineteen solar years is 6,939.60 days; 235 lunations is 6,939.69 days. The Metonic cycle may be written as 19 x 18.618 x 19.618 days.

YEARS	LUNATIONS	ERROR	EXAMPLE
☽☉ CALENDAR REPEAT CYCLES ☉☽			
3	37	3 days	
5	62	2 days	Coligny Calendar
8	99	1½ days	Avebury
19	235	2 hours	Stonehenge

☉ = 365.242 days   ☽ = 29.53059 days

BOSCAWEN-UN

THE PRECESSIONAL CYCLE
throwing a 26,000 year wobbly

The Earth's axis will not point to the present pole star forever. Like a spinning top that has tilted over a little and begun to wobble, or *precess*, the axis describes a complete circular rotation over about 25,920 years, tracing out the northern and southern circle of pole stars over this period. This is the *precessional cycle*, also known as a *Great Year* (*opposite*).

Equinoxes occur when the axial tilt of the Earth is at right angles to the Sun rather than facing toward or away from it (*see page 9*). This tilt is itself rotating backward very slowly, and the stars behind the Sun, on any given day of the year, change very slowly over time. The *Age of Pisces* commenced in A.D. 1. The *Age of Aquarius*, the next *Great Month*, will commence about A.D. 2160 when, at spring Equinox, the stars behind the Sun are those of the constellation of Aquarius. Magically, the diameter of the Moon is 2,160 miles, evoking the 2160 year length of the Great *Moonth*. Space becomes time!

In a human lifetime, precession is experienced as a single degree change in the position of the Sun against the fixed stars on a given date in the year. The precessional effect is caused by orbital asymmetries of the Sun and Moon.

The Earth's axial tilt itself varies between 21.5° and 24.8°, taking 41,000 years to complete a cycle.

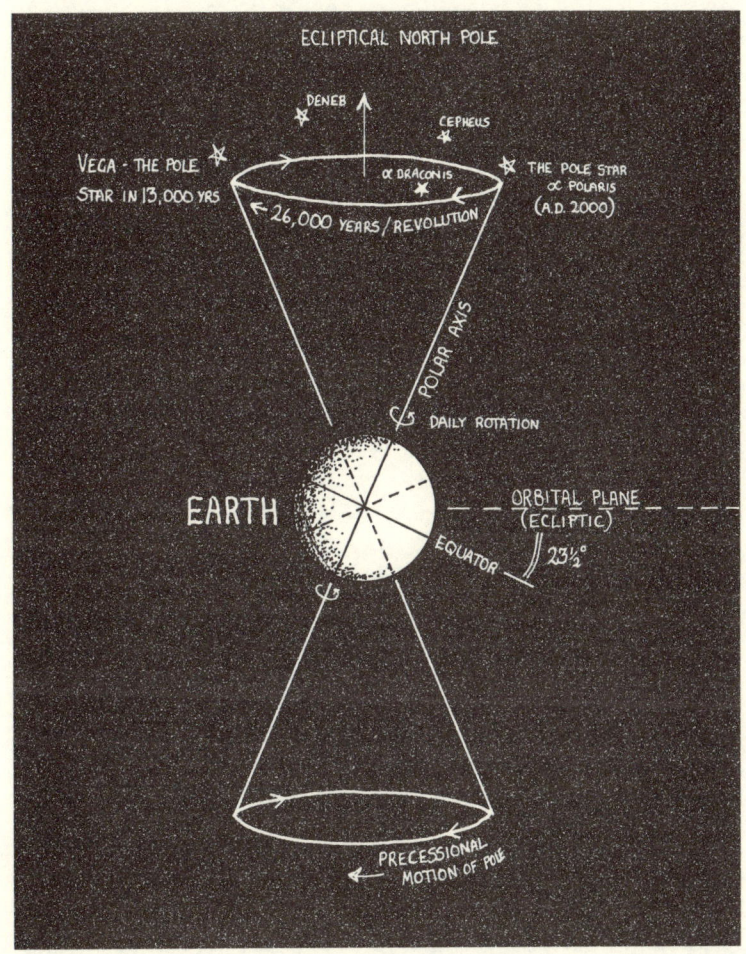

THE SOLAR YEAR
packets of days, leap years, and marking time

The solar year is 365.242 days long, practically 365. People who haven't ever thought the matter through will often tell you there are 365 "and a quarter" days in the year, but one cannot ever experience a quarter day in isolation. Days come in packets of one, and 365 of these make up the year, except that every fourth year an extra day slips in to make it 366.

At high latitudes, consecutive sunrises around the equinoxes are spaced more than the Sun's diameter apart (*five sunrises shown opposite, top*). However, each year the vernal equinox sunrise will appear from a slightly different position on the horizon—about one quarter of a degree (*opposite, bottom*). During three years of observation, the Sun appears to rise to the left of the original alignment until, in the fourth *leap* year, it rises once more very close to the original position, the tally for the year becoming 366 days. This accounts for the "quarter day" and is the basis for the additional intercalary day, February 29.

Over longer time periods than four years one gets the chance to obtain the length of the year with even more accuracy, by observing certain key years when the Sun rises precisely behind a foresight, stone marker, or notch in a distant mountain peak, a perfect repeat solar cycle.

The best of these occurs after 33 years, 12,053 days or sunrises. This is a staggeringly accurate repeat cycle and, remarkably, seems to have been known since prehistoric times.

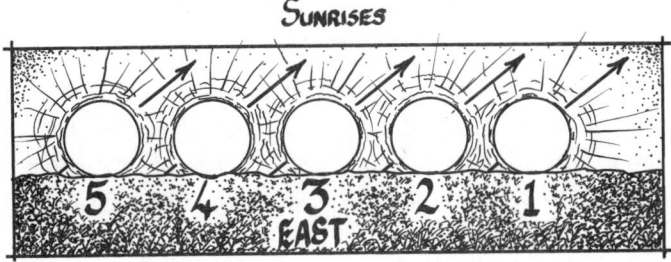

Sunrises

5   4   3   2   1

EAST

YEAR 3
1
START

EAST

Lat 55°N

THIRTY-THREE
the number of the solar hero

Thirty-three is a significant calendar number, which threads its secret through human culture. At a neolithic equinoctial alignment in Scotland, a cache of 33 tightly packed quartz pebbles was discovered by archaeologists. A renowned Irish megalithic site contains a stone with 33 chevrons picked out, and another has a snake motif with 33 folds (*opposite, bottom*). Ancient stories about the heroic *Tuatha de Danaan* frequently use the number 33. The first battle of *Mag Tuired* was fought by a saviour-hero *Lug* and 32 other leaders. In the second battle 33 leaders of the *Fomore* perish, 32 plus their highest king.

The Christian "highest king," Jesus, was crucified and resurrected at 33 years of age, rising again from behind a large stone. Islamic and Jewish calendar traditions recognize that it takes 33 solar years to complete 34 lunar years (of 12 lunations). A lunar calendar therefore cycles around the seasons, *Ramadan* falling 11 days earlier each year. The Masonic Order recognizes 33 degrees of proficiency, while the lonely game of Solitaire (*opposite, top*) consists of 33 holes and 32 balls.

These cultural artifacts share a common numerical factor, apparently derived from long-term observations of the Sun in prehistory. Pairing this cycle of the Sun with the nodal period of the Moon we may now "square the circle" by *area*, thereby solving the "third greatest problem of antiquity." Circle and square have the same area, a marriage of Heaven and Earth.

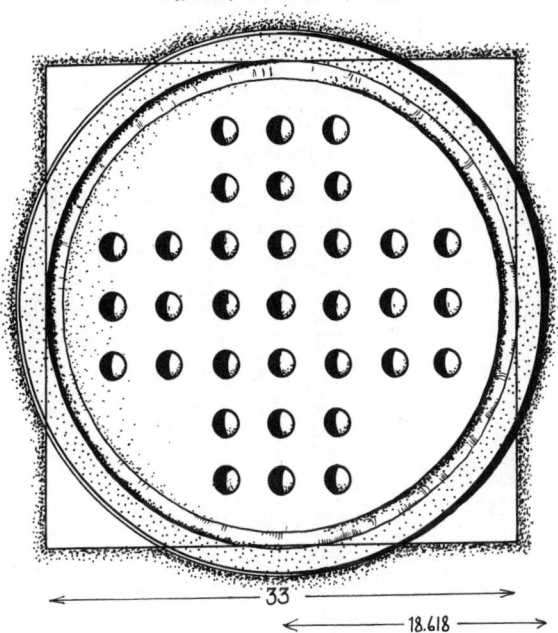

~ SQUARING the CIRCLE by AREA ~

33

18.618

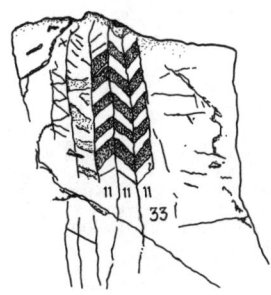

11 11 11
33

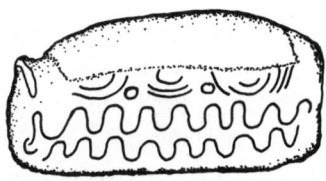

Early Irish Petroglyphs

41

DESIGNING A CALENDAR
13 months of 28 days

Timekeeping naturally starts with a calendar, which ideally should conform both to the solar year of 365 days *and* the Moon's phases. If the calendar year also divided easily into weeks and seasons and showed the phase of the Moon, the tides, and when to expect eclipses, we would be well pleased.

Starting from scratch, a calendar designer would soon discover that the Moon passes the same star in the sky every 27.3 days while the Sun takes 365 days. Dividing one by the other and choosing the nearest whole number, our designer would soon settle for 13 months of 28 days in a year. This gives a calendar year of 364 days—a number divisible by 2, 4, 7 and 13—a 52-week year with four 13-week seasons each of 91 days, a year with 13 months. All whole numbers and every year a leap year!

Practically, our designer could arrange 28 markers around the perimeter of a circle and arrange for a "moon-pole" to be moved counterclockwise once a day. A "sun-pole" would then be moved in the same direction, only thirteen times more slowly. This is the simplest way of providing a practical *soli-lunar* calendar, only requiring occasional resetting of the "Moon" (place the moon-pole opposite the sun-pole when full). If the stars were then represented on the circumference of the circle, the date, season, state of the tides (*see page 16*), lunar position, and lunar phase could be read off at a glance. We would gladly pay our designer.

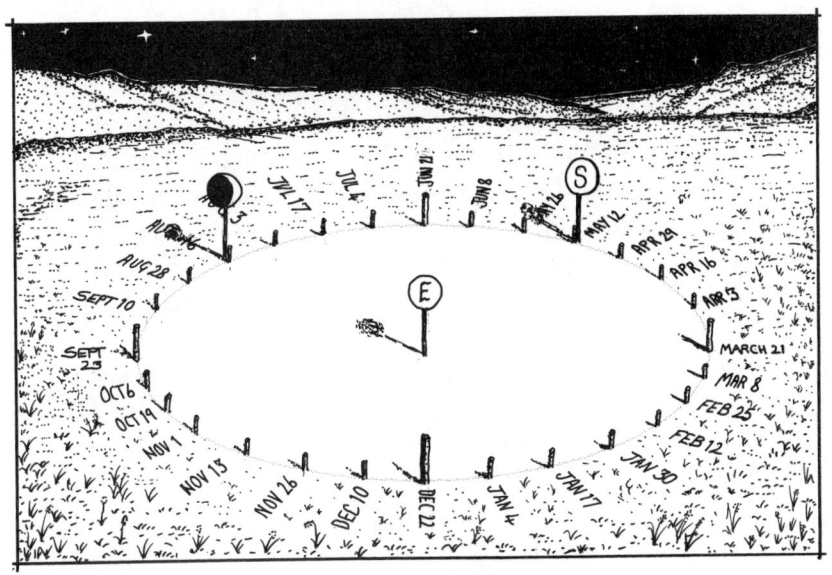

The model above shows a waxing quarter moon on May 12. Twenty-eight separate and equally spaced postholes indicate the Moon's angular motion per day—about 13°. Moving a "moon-pole" counterclockwise one hole per day emulates what happens in the sky, and the Moon takes 28 days to make a circuit. A "sun-pole" is now moved one hole every 13 days, taking 364 days to make the same journey, this being the best approximation to astronomic truth with the fewest holes (97.5 percent for the Moon and 99.66 percent for the Sun). It is therefore the basis for the ancient 13 month, 364 day calendar. The improved version (page 45) also predicts eclipses and is 99.9 percent accurate for the Moon, 99.8 percent for the Sun.

STONEHENGE
the oldest known calendar

The model shown on the previous page may be improved to include eclipse prediction. By doubling the number of postholes to 56 we exploit the useful coincidence of there being almost exactly 3 cycles of the lunar nodes in twice 28 years. The lunar nodes are here shown as two diametrically opposed triangular markers, initally placed near to the date-positions where eclipses have been recently observed.

The sun-pole now moves 2 holes counterclockwise every 13 days, and the moon-pole 2 holes counterclockwise every day. The node marker is moved three holes *clockwise* every year. When the sun-pole lies within three holes of a "node-pole," an eclipse of some sort will take place at each full or new moon, although it may not be visible at the location. If at each new moon the moon-pole is made to "skip" past the sun-pole (omitting this hole), this calendar will run for at least a year before the moon-pole needs resetting.

The design is shown opposite, hidden within the plan of Stonehenge, and indicates the date (season), Moon position, and phase, and predicts full and new moons and eclipses. The Aubrey circle (3000 B.C.) comprises 56 holes, accurately arranged round the perimeter of a 283-foot-diameter circle, predating the erection of the famous massive inner Sarsen circle. Perhaps posts were once placed in these holes; now they are filled with concrete, but this oldest known calendar would still work today.

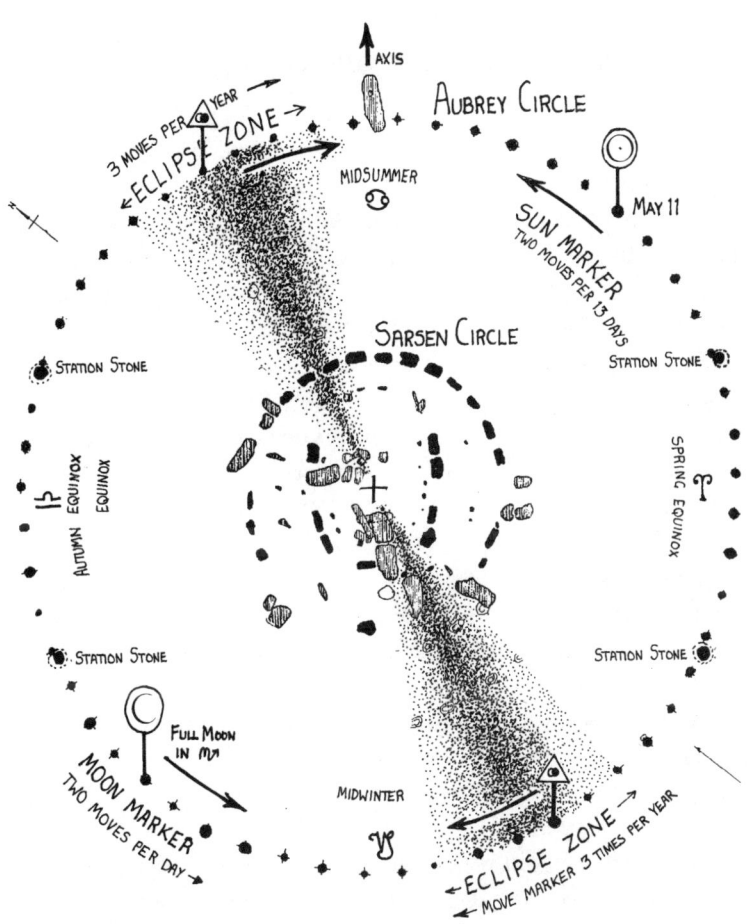

AXIS

3 MOVES PER YEAR →

← ECLIPSE ZONE →

AUBREY CIRCLE

MIDSUMMER

SUN MARKER
TWO MOVES PER 13 DAYS

MAY 11

SARSEN CIRCLE

STATION STONE

STATION STONE

SPRING ♈ EQUINOX

AUTUMN ♎ EQUINOX

EQUINOX

STATION STONE

STATION STONE

FULL MOON
IN ♏

MOON MARKER
TWO MOVES PER DAY →

MIDWINTER

♑

ECLIPSE ZONE →

← MOVE MARKER 3 TIMES PER YEAR →

45

TIME AND TIDE
making a calendar and tidal predictor

The instrument depicted opposite is a practical calendar and eclipse and tidal predictor based on the Stonehenge Aubrey circle. It may be built and used by anyone wishing to become more aware of the rhythms of the cosmos. It is easy to learn to set up and maintain this device. You can then know the height of the tide *before* setting off for the beach!

Place the Sun marker at the current date and use an almanac to determine the Moon's position, or wait until a new or full moon. If the Moon is visible, its position can often be set approximately from its phase. The nodal axis (eclipse zone) moves clockwise three markers a year (correctly set for January 1, 2001).

The Moon's position in the sky at high tide needs first to be known for the chosen location. This varies from country to country, and to set this position, "dawn" and "dusk" are used to indicate the local horizon. The quadrant arms are then clamped at the correct angle onto the rotating central 24-hour clockface.

From now on rotate the central clock (with clamped quadrant arms) until one of the "High Tide" markers points to the current Moon position. Simply read off the time of high tide from the clock time *adjacent to the current Sun position*. For low tide, point a "Low Tide" arm and repeat the same procedure. Move the Sun and Moon every day (*as on previous page*). Big spring tides follow full and new moons, small neap tides follow the quarter moons.

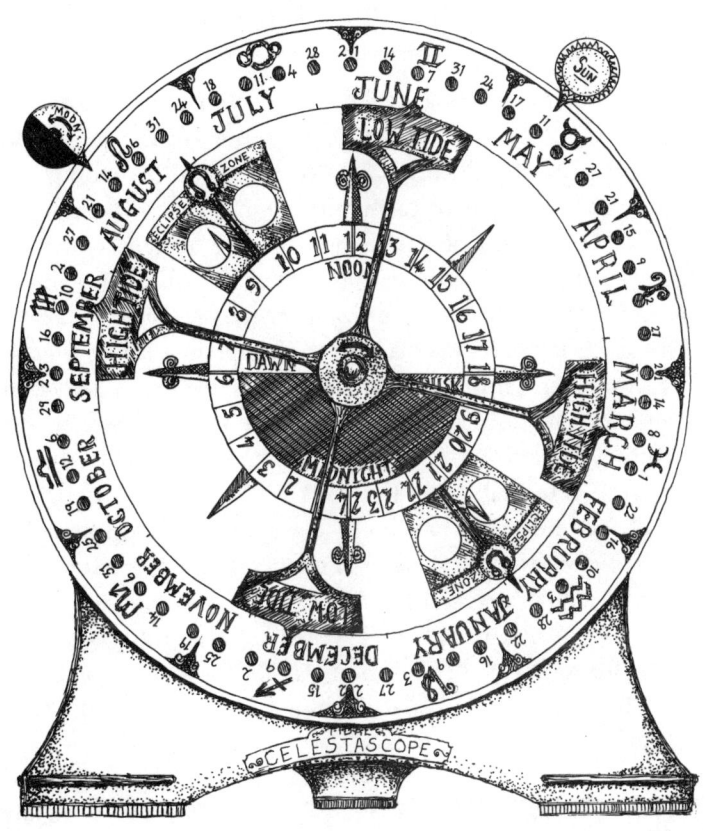

THE SILVER FRACTION
between 12 and 13 full moons

The secret of the calendar is the 11-day mismatch between the lunar year (12 lunations) and that of the solar year (12.368 lunations). Enoch called this the "over-plus of the Moon," but more poetically we may call it the *silver fraction*. Remarkably, three common units of length used by the ancient world, the *Megalithic yard*, the *Royal cubit,* and the *foot*, relate through the astronomy of the lunation and the silver fraction (*opposite, top*).

The silver fraction is actually 10.875 days in length, which is 0.368 lunations; almost exactly seven-nineteenths as a fraction. Intriguingly, the diameters of the two main circles of Stonehenge, the Aubrey circle (283 feet) and the Sarsen circle (104 feet) are in the ratio 7:19 to each other (*below*).

Simple geometry can also reveal the annual lunation figure. A pentagram drawn inside a circle of diameter 13 units has star arms of length 12.364 (*opposite, bottom*)! All 5 star arms add up to 61.82, the number of full moons in 5 years and also 100/*phi* (*to 99.9 percent—see page 28*). The famous Celtic *Coligny* calendar (100 B.C.) is based on a 5-year, 62-lunation cycle (*see page 35*). *Phi* in the sky!

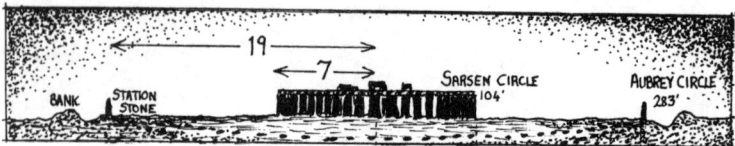

48

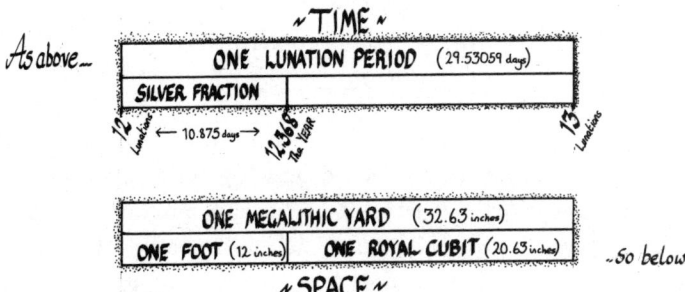

As above...

~TIME~

ONE LUNATION PERIOD (29.53059 days)	
SILVER FRACTION	

12 Lunations ← 10.875 days → 12.368 The YEAR   13 Lunations

ONE MEGALITHIC YARD (32.63 inches)	
ONE FOOT (12 inches)	ONE ROYAL CUBIT (20.63 inches)

~SPACE~   *~So below.*

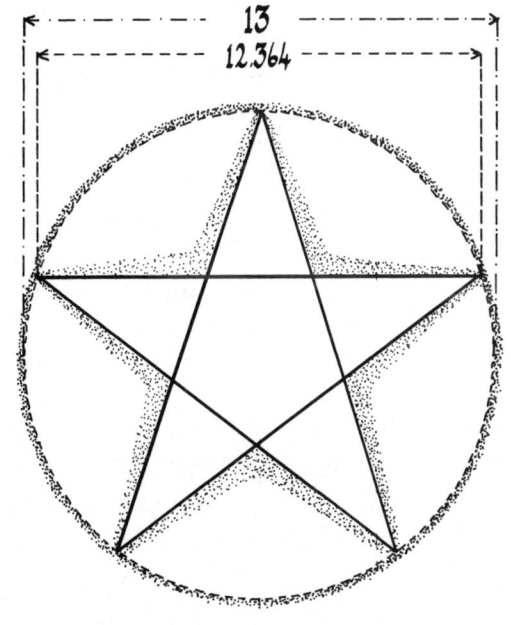

13
12.364

THE LUNATION TRIANGLE
Solomon meets Pythagoras

A 5:12 rectangle has a diagonal length of 13. The four station stones at Stonehenge define one (*see page 45*), as do the proportions of Solomon's Temple at Jerusalem. Twelve plus a thirteenth, as the redeeming Savior, occurs in many heroic stories, Jesus, King Arthur, and the Mayan wind god Kukulcan being examples. To the Pythagoreans the number 5 signified completeness or marriage, formed as the first male number, 3, becomes wedded to the first female number, 2.

The true number of months in the year falls between 12 and 13, and in order to define a true soli-lunar calendar this figure, 12.368, must be determined. The *lunation triangle* is defined as a 5:12:13 right triangle, the second *Pythagorean triangle*, with the "5" side divided as 3:2. A new hypotenuse to this point measures 12.369. The Moon, 13, thereby becomes married to the Sun, 12, where the female, 2, joins to the male, 3. The sacred marriage of Sun and Moon, made in Heaven, is witnessed on Earth, and occurs at the musical fifth, the most harmonious interval (3:2). Musical allegories abound (*opposite, bottom*), and Solomon's throne is wisely placed at the 3:2 point in the Temple.

St John's Gospel ends with a fishy story. Jesus reappears for the third time since his resurrection and instructs his fishless disciples "cast your nets on the right side," who then catch 153 fish.

The square of the annual lunation rate is 153—12.368—to 99.99 percent.

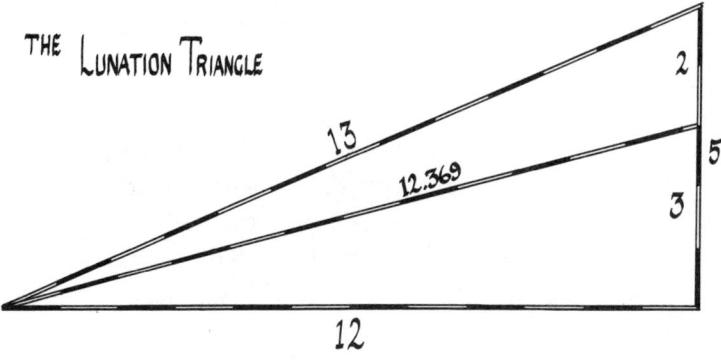

THE LUNATION TRIANGLE

2

13

5

12.369

3

12

~ the Keyboard ~

OCTAVE

The Chromatic Scale
12 notes plus a completing 13th (the octave)
5 "black" notes, arranged as 3 and 2

SUN, MOON, AND EARTH
the revealed structure of the system

The diagram opposite frames a curious symmetry. Both the eclipse year (346.62 days) and thirteen full moons (383.89 days) are almost exactly equally spaced, at 18.6 days, on either side of the solar year of 365.242 days. Because the eclipse year is the square of the lunar node period in days (18.618^2), we are now able to write that the solar year is (18.618 x 18.618) + 18.618. This is also 18.618 x 19.618. The number 18 added to $1/phi$, *phi*, or phi^2, now delivers the following extraordinary formulas to 99.99 percent:

18.618 x 18.618 = 346.62 days (the eclipse year)

18.618 x 19.618 = 365.242 days (the solar year)

18.618 x 20.618 = 383.89 days (13 lunations)

The astronomy reveals the actual geometry and numerical structure of the Sun, Moon, Earth system (*opposite*). Imagine a solar eclipse at (1). The Sun then moves to meet up with the same node after an eclipse year (2), 346.62 days later. Passing the original eclipse point (1) at the end of one year, the Sun and Moon then meet for the thirteenth lunation at (3).

An isosceles triangle, drawn to fit the angles generated by the astronomy, also then defines the solar eclipse limits, and has the remarkable property of replicating the numbers shown above as ratios. In addition, the shorter side has a length 12.368, the annual lunation rate. One marvels at this revelation of cosmic order!

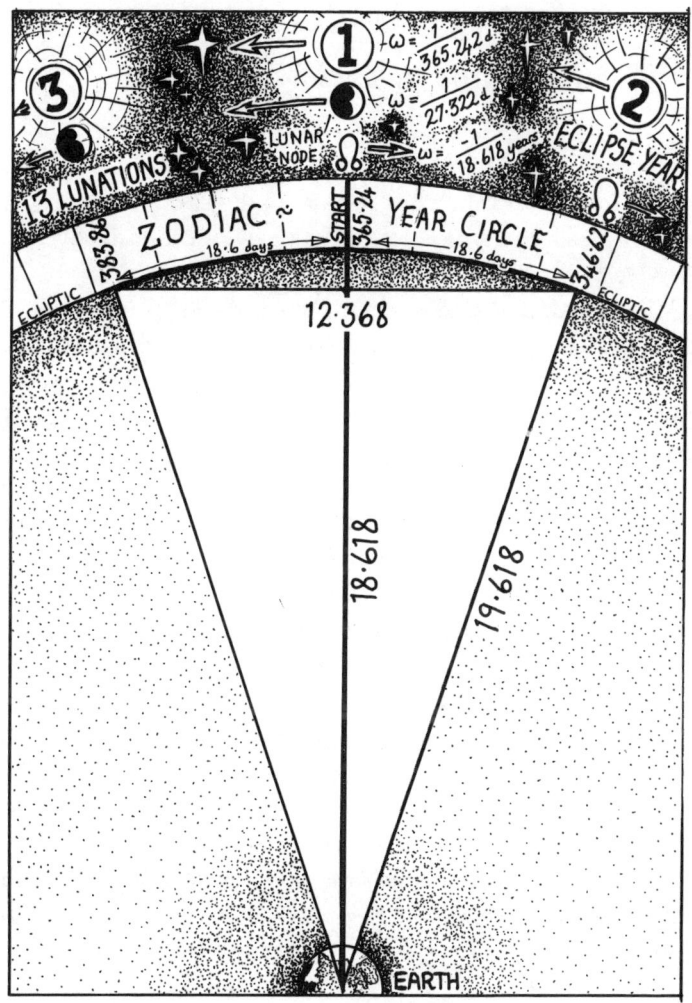

① $\omega = \dfrac{1}{365.24.2\,d}$

$\omega = \dfrac{1}{27.322\,d}$

LUNAR NODE ☊ $\omega = \dfrac{-1}{18.618\ \text{years}}$

③ 13 LUNATIONS

② ECLIPSE YEAR ☋

ZODIAC ♈   START   YEAR CIRCLE   ☋

383.86   18.6 days   365.24   18.6 days   346.62

ECLIPTIC   ECLIPTIC

12.368

18.618

19.618

EARTH

A STONE-AGE COMPUTER
neolithic cosmology revealed

In remote places in Britain survive many examples of curiously flattened stone circles, constructed some 4,500 years ago and named *type-A* and *type-B* by Professor Thom, the discoverer of the Megalithic yard (2.72 feet, 32.64 inches, 0.83 m).

Both type-A and type-B rings invoke a Christian symbol, the *vesica piscis*, the almond shape between two overlapping circles, here applied 2,500 years before Jesus. Stones are commonly placed intelligently to the geometry (*opposite*). More astonishingly, this design invokes the same triangle we have just witnessed underpinning the structure of the Sun, Moon, Earth system!

The right triangle has side ratios $1:3:\sqrt{10}$. A rope taken from the center, O, to point P, and thence to B and A has a length $3 + \sqrt{10} + 2$, which is 8.16227, exactly one quarter of the Megalithic yard, in inches. If length PB is considered to represent a lunation period, then the intersection of the vesica circle cuts it at 0.368 of its length (at X). If PB is now considered to represent the solar year, then PO represents the eclipse year, this ratio being either $\sqrt{10}:3$ or 19.618:18.618. The ratios foot:Royal cubit:Megalithic yard may also be "read" from this exquisite device.

This most beautiful analogue of the Sun, Moon, Earth system stores their key constants *and* the ancient metrology all within itself, as ratios. An awesome glimpse of an ancient wisdom is now finally revealed.

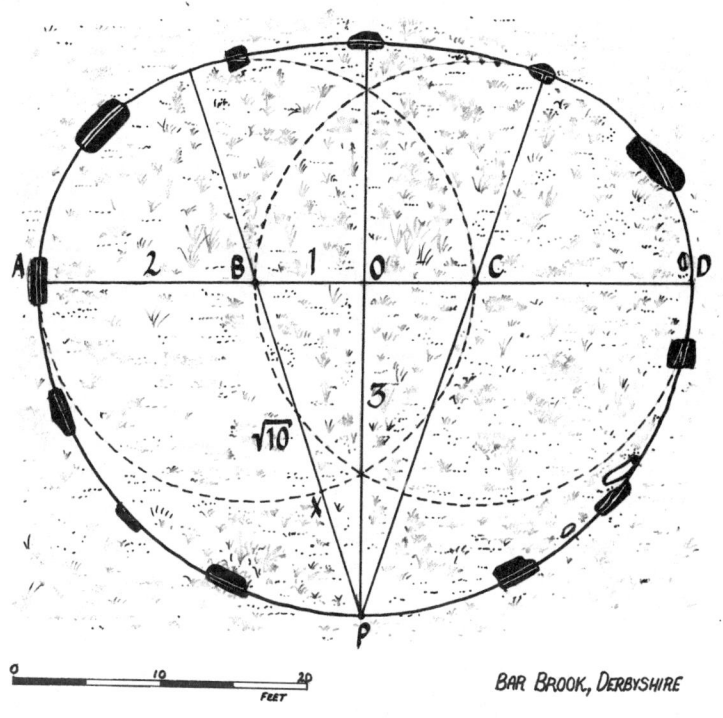

Labels within the figure:
A 2 B 1 O C D
3
√10
√10
X
P

BAR BROOK, DERBYSHIRE

0   10   20
        FEET

Foot: Royal Cubit: Megalithic Yard = 1 : 1.72 : 2.72

The Megalithic Yard is 2.72ft, 32.64" (4 × 8.1623")
The Royal Cubit is 1.72ft, 20.64"

$$\frac{\sqrt{10}}{\sqrt{10}-2} = 2.72 \qquad \frac{\sqrt{10}}{\sqrt{10}-5} = 1.72 \qquad 2 + 3 + \sqrt{10} = 8.1623$$

55

PEBBLES ON THE SHORES OF TIME
multidimensional solutions to local cosmology

18:618 days
(ONE "NODE-DAY")

18.618 years
MOON'S NODAL PERIOD

18.618

AREA = 346.63
✕ ECLIPSE YEAR ✕
(DAYS)

18.618

ECLIPSE YEAR = 18.618² DAYS

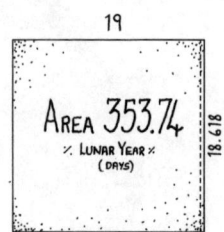

19

AREA 353.74
✕ LUNAR YEAR ✕
(DAYS)

18.618

12 LUNATIONS = 18.618 × 19

19.618

AREA = 365.247
✕ THE YEAR ✕
(DAYS)

18.618

SOLAR YEAR = 18.618 × 19.618 DAYS

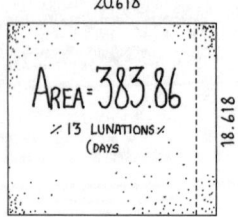

20.618

AREA = 383.86
✕ 13 LUNATIONS ✕
(DAYS

18.618

13 LUNATIONS = 18.618 × 20.618 DAYS

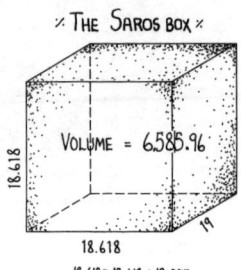

✕ THE SAROS BOX ✕

VOLUME = 6,585.96

18.618

18.618

19

18.618 × 18.618 × 19 DAYS

✕ THE DRAGON'S BOX ✕

VOLUME = 6800.2

18.618

18.618

19.618

18.618 × 18.618 × 19.618 DAYS

✕ THE METONIC BOX ✕

VOLUME = 6,939.78

18.618

19.618

19

18.618 × 19.618 × 19 DAYS

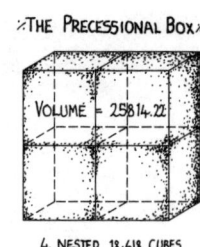

✕ THE PRECESSIONAL BOX ✕

VOLUME = 25814.22

4 NESTED 18.618 CUBES
= 4 × 18.618³ YEARS

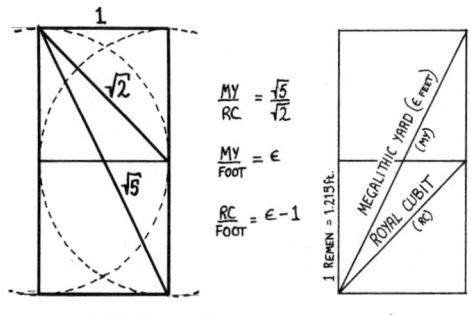

$$\frac{MY}{RC} = \frac{\sqrt{5}}{\sqrt{2}}$$

$$\frac{MY}{FOOT} = \epsilon$$

$$\frac{RC}{FOOT} = \epsilon - 1$$

The Remen: An Egyptian unit of 1.215 ft

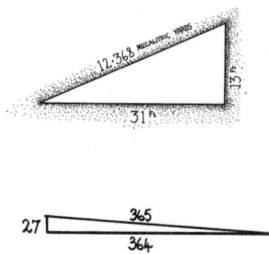

∾ CALENDAR TRIANGLE ∾

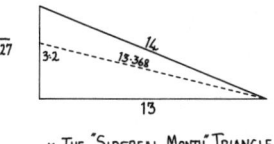

× THE "SIDEREAL MONTH" TRIANGLE ×

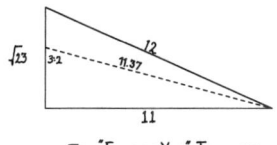

× THE "ECLIPSE YEAR" TRIANGLE ×

× THE "SAROS - METONIC" TRIANGLE ×

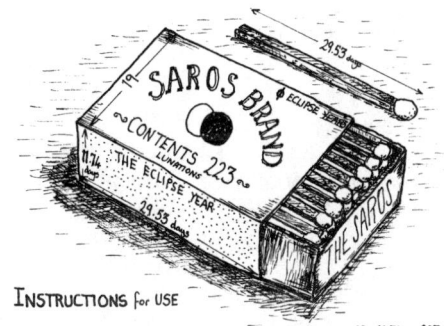

INSTRUCTIONS for USE

1. OPEN SAROS DRAWER.

2. REMOVE A LUNATION MATCH

3. STRIKE AGAINST THE ECLIPSE YEAR
   AND THEN...

4. SET FIRE TO ALL PREVIOUS
   IDEAS YOU HAD ABOUT THE MOON.

$19 \times 11.74 = 223$

$19 \div 11.74 = \phi$

$19 \times 29.53 = \phi$ ECLIPSE YEARS

VOLUME
$= 29.53 \times 11.74 \times 19$
$= 6,585.97$ days
$= $ SAROS

TIMES

Solar Year: 365.242199 days
Lunar Year: 354.367 days
Lunation period: 29.53059 days
Eclipse Year: 346.62 days
Lunation rate: 12.36826623/year
Saros eclipse cycle: 18.03 years or 223 lunations or 6585.322 days
Lunar node cycle (Draconic year): 18.618 years or 6800.0 days
Metonic cycle: 19 years or 235 lunations or 6939.602 days

Sidereal lunar month: 27.322 days
Sun-spot cycle: 11 years
Sidereal day: 23 hours, 56 minutes, and 4 seconds
Solar tropical day (clock time): 24 hours
Lunar day (average): 24 hours, 52 minutes, and 4.31 seconds
Precessional cycle: approx 25,820 years
Tidal spacing (average): 12 hours, 26 minutes, and 2.15 seconds

LENGTHS

One foot: one degree of arc along the equator ÷ 365,242
Megalithic yard: 2.72 feet (+/- 0.003 feet).
Earth's equatorial radius: 3963.4 miles
Polar radius: 3950.0 miles

Lunar radius: 1,080 miles
Lunar distance: 222,000–253,000 miles mean: 240,000 miles
Sun's radius: 432,000 miles
Sun's distance (mean): 93,009,000 miles

ANGLES & RATIOS

Lunar orbit inclination to Earth-Sun plane: 5° 8' 30"
Solar angular diameter (mean): 0° 32'
Lunar angular diameter (mean): 0° 31' 30"

Earth's axial tilt: 23° 27'
Earth to Moon density ratio: 1.6 : 1
Earth to Moon mass ratio: 81:1
Eccentricity of lunar orbit: 1/18

SOME DEFINITIONS

Lunation period: Time between consecutive new moons (or full moons).
Ecliptic: The Sun's apparent path through the zodiacal belt of stars, seen from Earth.
Solar tropical year: Time between consecutive sping equinoxes.
Precessional (Great) year: Time for the zodiac to rotate (backward) around the calendar.
Syzygy: Sun, Moon, and Earth in a line.

Solar/Lunar day: Time between consecutive south transits of the Sun/Moon.
Lunar year: 12 lunations.
Sidereal lunar period: Time for the Moon to return to the same longitude (or star).
Lunar nodes: Two opposite points where the Moon's path crosses the ecliptic.
Eclipse year: Time between consecutive conjunctions of the Sun and north node.

IRRATIONALS

e = 2.718282 . . .
pi(π) = 3.141593 . . .

√2 = 1.414214 . . .
√3 = 1.732051 . . .

phi(ø) = 1.618034 . . .
√5 = 2.236068 . . .